QCE

Biology

Revision Workbook

* greenhill

Copyright © Gemma Dale 2024

All rights reserved. No part of this publication may be reproduced, distributed, or transmitted in any form or by any means, including photocopying, recording, or other electronic or mechanical methods, without the prior written permission of the publisher, except in the case of brief quotations embodied in critical reviews and certain other non-commercial uses permitted by copyright law.

✳ greenhill

https://greenhillpublishing.com.au/

Dale, Gemma (author)

QCE Biology: Revision Workbook

ISBN 978-1-923333-46-8

TEXTBOOK | BIOLOGY

Typesetting Forma DJR 10.5/16

Cover and book design by Green Hill Publishing

Diagrams by Matilda Crossley

QCE

Biology

Revision Workbook

DR GEMMA DALE

ILLUSTRATED BY
MATILDA CROSSLEY

CONTENTS

Unit 1

Introduction — 1

Cells and multicellular organisms — 5

Topic 1: Cells as the basis of life — 6
Section 1: Cell Structure — 6
Section 2: Cell Division — 11
Section 3: Cell Membrane — 16
Section 4: Cell Transport — 19

Topic 2: Exchange of nutrients and wastes — 23
Section 1: Biochemistry — 23
Section 2: Enzymes — 28
Section 3: Exchange Surfaces — 35
Section 4: Excretory System — 39

Topic 3: Cellular energy, gas exchange and plant physiology — 42
Section 1: Respiration — 42
Section 2: Photosynthesis — 46
Section 3: Gas Exchange — 48
Section 4: Plant Physiology — 51

Unit 2

Maintaining the internal environment — 55

Topic 1: Homeostasis - thermoregulation and osmoregulation — 56
Section 1: Nervous System — 56
Section 2: Endocrine System — 62
Section 3: Thermoregulation — 66
Section 4: Osmoregulation — 69

Topic 2: Infectious disease and epidemiology — 74
Section 1: Disease — 74
Section 2: Immune Response — 77
Section 3: Epidemiology — 86

Intro

Introduction

Welcome to your QCE Biology Revision Student Workbook for Units 1 and 2. This workbook is designed to be your companion throughout your lessons or to be completed for homework, helping you to consolidate and deepen your understanding of the material covered.

In Unit 1 and 2, you will explore key concepts and develop essential skills that are foundational to your biology studies. Each topic has been split into different sections, providing a structured approach to your learning.

Unit 1: Cells and multicellular organisms	Unit 2: Maintaining the internal environment
Topic 1: Cells as the basis of life	Topic 1: Homeostasis – thermoregulation & osmoregulation
Topic 2: Exchange of nutrients and waste	Topic 2: Infectious disease and epidemiology
Topic 3: Cellular energy, gas exchange and plant physiology	

The workbook has been designed to reinforce what you've learned in class and to provide you with additional practice opportunities. You'll find a range of activities, all aimed at helping you master the content and apply your knowledge effectively. At the start of each section, you'll find a Self-Assessment Table designed to help you gauge your confidence in the syllabus content. This table allows you to reflect on your current understanding and track your progress as you work through the section.

Use this workbook as a tool to review and reinforce your learning, track your progress and prepare for assessments. By engaging with the material actively and consistently, you'll build a strong foundation for success in your studies.

Unit 1

Cells and multicellular organisms

TOPIC 1: CELLS AS THE BASIS OF LIFE

Section 1: Cell Structure

I can....	Confidently	Somewhat	Not quite
...compare prokaryotic and eukaryotic cells			
...identify key organelles and their functions			

Question 1

Most eukaryotic cells contain mitochondria, which are comparable in size to prokaryotic cells and share several characteristics with them. According to the endosymbiotic theory, these ancient prokaryotes were engulfed by other bacterial cells, forming a mutually beneficial relationship that eventually led to the development of eukaryotic cells. Identify two structures present in both prokaryotic cells and mitochondria. (2)

____

Question 2

Describe two ways in which mitochondria differ from prokaryotic cells. (2)

____

____

Question 3

The electron micrographs below show organelles found in eukaryotic cells.

A   B   C

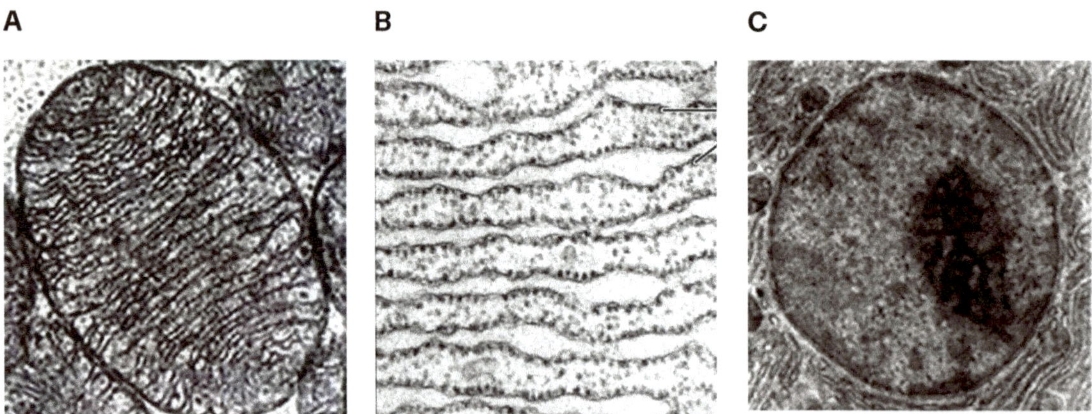

Identify the organelles in photographs A, B and C and state their function. (6)

A ____

B ____

C ____

Question 4

Complete the table to show the ways prokaryotic and eukaryotic cells differ in their structure. (3)

Prokaryotic cells	Eukaryotic cells

Question 5

Complete the table by putting a tick (✓) which cell/s the feature is found in. (5)

Feature	Prokaryotic cell	Eukaryotic cell
Cell membrane		
Nucleus		
Ribosomes		
Rough endoplasmic reticulum		
Mitochondria		

Question 6

Name two structures found in animal cells that are absent in plant cells. (2)

____

____

Question 7

State two functions of a vacuole. (2)

____

Question 8

Explain the function of lysosomes. (2)

Question 9

State two organelles that contain DNA. (2)

Question 10

Mucus consists of water, carbohydrates, proteins and triglycerides. It is secreted by goblet cells. The figure shows a goblet cell.

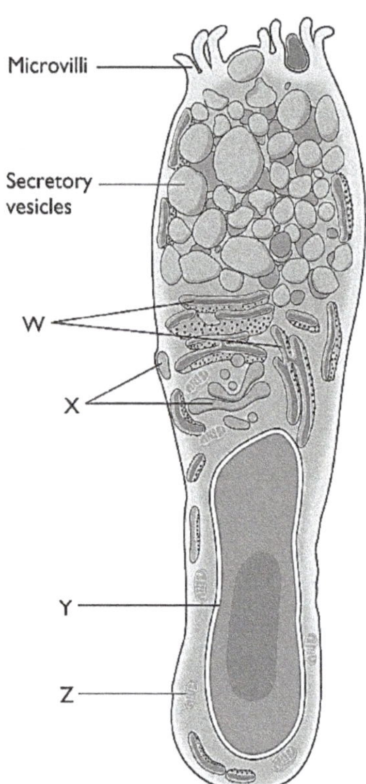

a) Name W, X, Y and Z. (4)

W ______________________________

X ______________________________

Y ______________________________

Z ______________________________

b) Propose why goblet cells contain a high number of organelle Z. (1)

____

____

____

____

c) Explain how organelle X is relevant to the function of the goblet cell. (2)

____

____

____

____

Question 11

The diagram below shows a plant and animal cell.

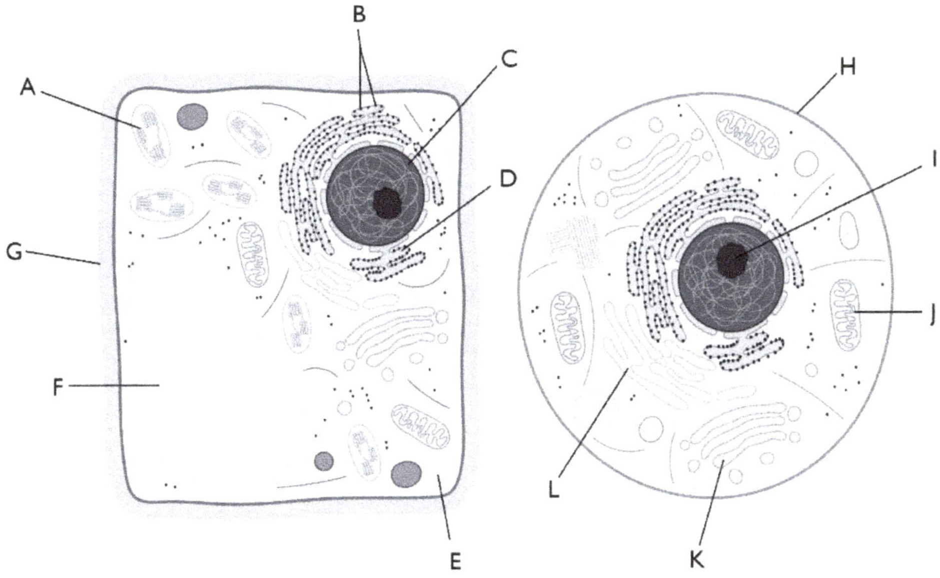

Complete the table below. (12)

Organelle	Name	Function
A		
B		
C		
D		
E		
F		
G		
H		
I		
J		
K		
L		

Section 2: Cell Division

I can....	Confidently	Somewhat	Not quite
...describe how stem cells originate through the process of mitosis and differentiate into specialised cells to form tissues			
...distinguish between unipotent, multipotent, pluripotent and totipotent stem cells			
...describe how the hierarchical organisation of cells, tissues, organs and systems allow multicellular organisms to obtain nutrients, exchange gases and remove wastes			
...explain that each body system contains specialised cells and tissues that are structurally suited to function			

Question 1

State what is meant by the term stem cell. (2)

____

Question 2

Fertilisation in humans occurs when a sperm cell fuses with an egg cell, leading to the formation of a fertilised cell, which then divides to form a cluster of totipotent cells. Describe what is meant by the term totipotent cell. (2)

____

____

Question 3

Stem cells have the potential to repair damaged organs. In the skin, for instance, stem cells can be used to repair severe wounds or burns. Explain why stem cells from the heart cannot be used to grow cells for repairing damaged skin. (2)

____

____

Question 4

Name the stage of the cell cycle where DNA replicates. (1)

____

Question 5

In humans, various types of stem cells are present in the bone marrow. Mesenchymal stem cells can differentiate into several types of cells, such as bone cells, cartilage cells, muscle cells, and fat cells. The skeletal system includes bone tissue, muscle tissue and cartilage tissue. Explain how a tissue differs in structure from a system. (2)

____

____

Question 6

Name the stages of mitosis during which:

a) The chromatids separate and move toward the poles. (1)

____

b) The nuclear membrane reforms and cytokinesis begins. (1)

____

c) The chromosomes align along the equator. (1)

____

d) The chromosomes become visible and the spindle begins to form. (1)

____

Question 7

Explain why referring to interphase as a 'resting stage' is incorrect. (2)

____

____

Question 8

State two purposes of mitosis. (2)

____

____

Question 9

Name the stage of mitosis shown in the diagram. They are not in the correct order. (4)

Question 10

Explain what is meant by the term tissue. (2)

____

____

Question 11

Fill in the gaps:

In the first stage of mitosis, known as _________________, the chromosomes become visible.
Each chromosome is made up of two sister chromatids, joined at the ______________.
The nuclear______________ breaks down, spindle fibres form and the ________________
chromosomes line up at the ________________. During ______________, the sister chromatids
separate, and are pulled to opposite sides of the cell by the ________________ contracting.
Separate nuclei form and two new daughter cells are formed that are ______________
identical. (8)

Question 12

Liver cells perform various metabolic functions and require a continuous supply of energy.
Which organelle would you expect to find in large numbers in liver cells to provide this energy?
Explain your response. (2)

____

____

Question 13

State what is meant by the term organ. (1)

____

Question 14

A koala body cell contains 16 chromosomes. The table below shows the number of
chromosomes and the mass of DNA in different nuclei from the same koala.
Complete the table. (2)

	Number of chromosomes	Mass of DNA/arbitrary units
At prophase	16	
At telophase		5
From a sperm cell		

Question 15

Stem cells can be sourced from embryos and menstrual blood, but these two cell types have
distinct differences. Embryonic stem cells can be used to treat a broader range of disorders
compared to menstrual blood stem cells. Explain why this is the case. (2)

____

____

____

Question 16

The figure shows the cell cycle.

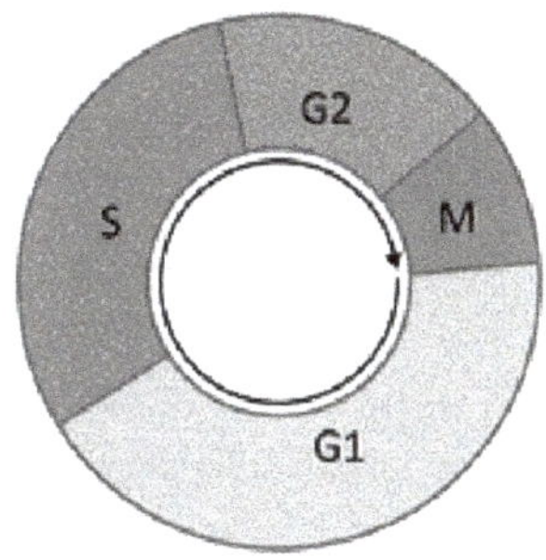

State what occurs during stages G1, S, G2 and M. (4)

Stage	Process
G1	
S	
G2	
M	

Question 17

The figure shows an image of onion root cells in different stages of mitosis. Annotate the figure to show identify cells undergoing interphase, prophase, metaphase, anaphase and telophase. Label one cell for each stage. (5)

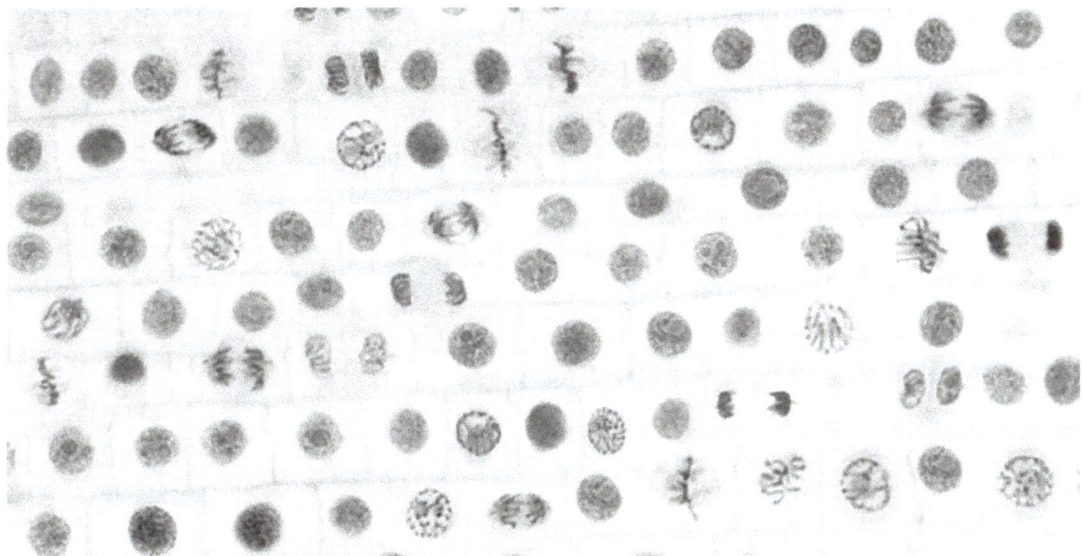

Section 3: Cell Membrane

I can....	Confidently	Somewhat	Not quite
...describe the structure and function of the cell membrane based on the fluid mosaic model			

Question 1

The diagram shows a plasma membrane.

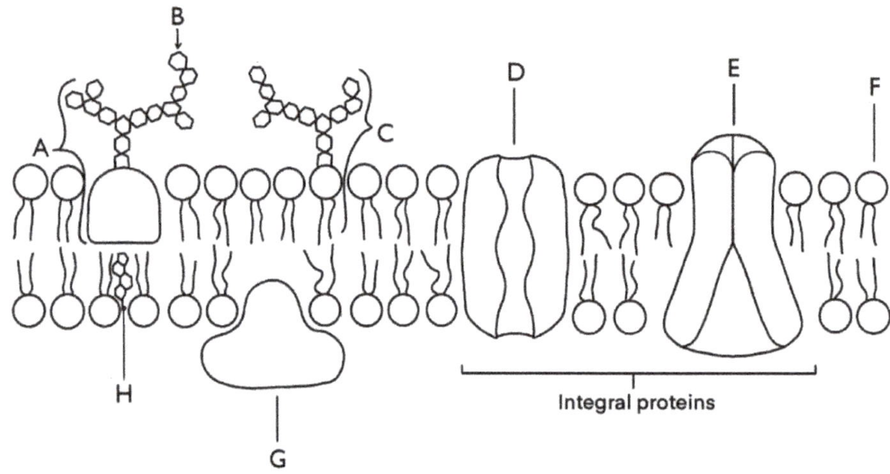

a) State the name given to the model of plasma membrane structure shown. (1)

____

b) Name molecules A to H.

A ____

B ____

C ____

D ____

E ____

F ____

G ____

H ____

Question 2

The phospholipid bilayer regulates the movement of molecules in and out of a cell. Explain why phospholipid molecules form a bilayer. (3)

____

____

____

____

Question 3

Describe the structure of a phospholipid. (2)

____

____

____

Question 4

Describe the function of cholesterol in cell membranes. (1)

____

____

____

____

Question 5

Beetroot vacuoles contain betacyanin, a red pigment. When beetroot discs are cut and placed in 70% ethanol (an organic solvent) at 20°C, the pigment leaks into the ethanol, turning it red.

a) Using your understanding of cell membrane structure, explain why the pigment leaks out. (2)

____

____

____

b) When the experiment was repeated at 30°C, the ethanol turned red more quickly. Explain why this occurred. (2)

____

____

____

Section 4: Cell Transport

I can....	Confidently	Somewhat	Not quite
...explain how the cell membrane regulates movement of substances into and out of the cell			
...compare active and passive transport			
...explain how the size of a cell is limited by surface area-to-volume ratio and rate of diffusion			
...interpret data from an experiment investigating the effect of surface area-to-volume ratio on the rate of diffusion			

Question 1

Active transport and facilitated diffusion are two methods in which substances cross plasma membranes.

a)   State one difference and one similarity between active transport and facilitated diffusion. (2)

Difference ____

____

Similarity ____

____

b)   Vitamins C and D are essential for proper body function and enter cells via different routes across the plasma membrane. Vitamin C is water-soluble, while Vitamin D is lipid-soluble. Explain how the structure of the plasma membrane influences the route taken by each vitamin. (4)

Vitamin C ____

____

Vitamin D ____

____

Question 2

Name a cell transport mechanism that does not require a concentration gradient to be present to take place. (1)

____

Question 3

Water crosses the plasma membrane by osmosis. Complete the table with ticks (✓) to show the following. (3)

a)   Show the direction in which water moves across the plasma membrane.

b)   If the cell will burst.

	Initial movement of water		Cell bursts?
	Into cell	Out of cell	
Plant cell in distilled water			
Animal cell in concentrated salt solution			
Animal cell in distilled water			

Question 4

The figure shows some cell organelles that are involved in the production of extracellular enzymes.

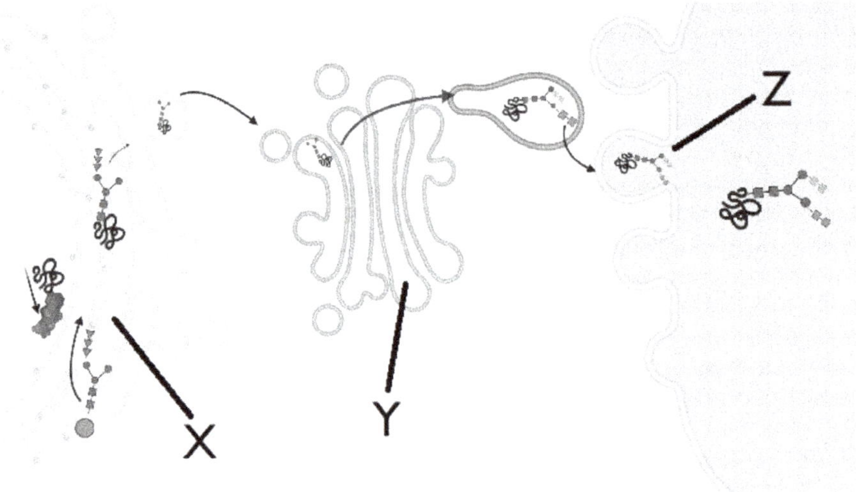

a)   Name organelles X and Y. (2)

X   ____

Y   ____

b)   Name process Z. (1)

Z   ____

c)   Describe the roles of X and Y in the formation and transport of extracellular enzymes. (3)

____

____

____

Question 5

Oxygen enters red blood cells through diffusion.

a)   Define diffusion. (1)

____

____

b)   Describe the adaptations of red blood cells that make them efficient at taking up and
transporting oxygen. (2)

____

____

____

____

Question 6

Model cells can be used to investigate how cell size affects the rate of diffusion. When agar cubes containing phenolphthalein indicator are placed in dilute hydrochloric acid, the time it takes for the cubes to become colourless can be used to measure the rate of diffusion.

a) Calculate the surface area to volume ratio for each of the cubes. (3)

Cube length (cm)	Surface area to volume ratio	Time to change colour (s)
1	_____ :1	145
2	_____ :1	365
3	_____ :1	712

b) Use your results to suggest why multicellular organisms require transport systems, but unicellular organisms do not. (2)

____

____

Question 7

Antigens on pathogen cell membranes can trigger endocytosis and exocytosis in white blood cells, both of which involve transport across the cell membrane. State two differences between endocytosis and exocytosis. (2)

____

____

____

TOPIC 2: EXCHANGE OF NUTRIENTS AND WASTES

Section 1: Biochemistry

I can….	Confidently	Somewhat	Not quite
…describe the structure and function of carbohydrates			
…describe the structure and function of lipids			
…describe the structure and function of proteins			

Question 1

a) State the components required to synthesise a triglyceride. (2)

____

____

b) Name the chemical reaction that joins these components. (1)

____

c) State one function of triglycerides in living organisms. (1)

____

Question 2

Describe the molecular structure of starch (amylose and amylopectin). (6)

____

____

____

Question 3

Lactose is a carbohydrate molecule made by joining a glucose molecule to a galactose molecule.

a) Name the bond that joins glucose and galactose in a lactose molecule. (1)

____

b) Name the type of reaction that breaks this bond. (1)

____

Question 4

State what group of biomolecules the following structures belong to. (6)

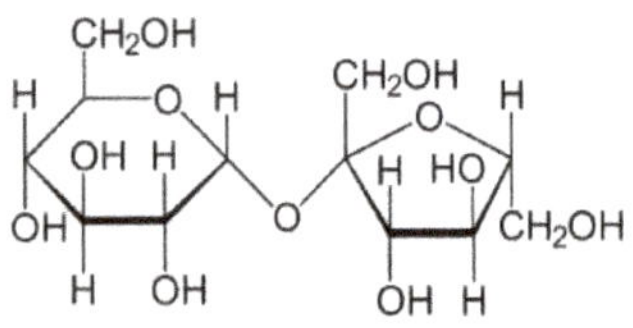	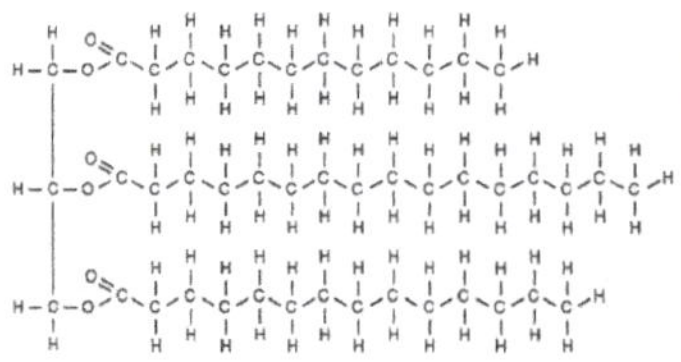	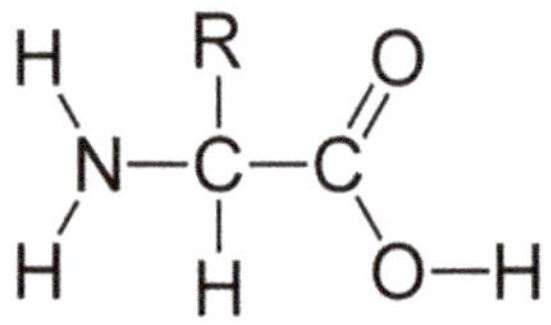
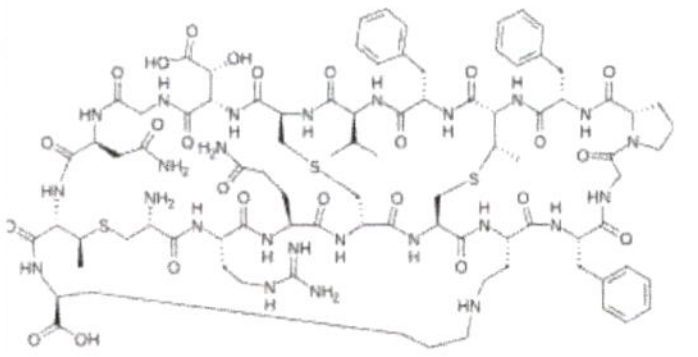	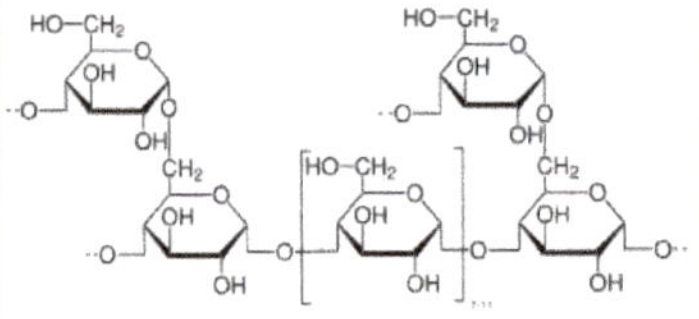	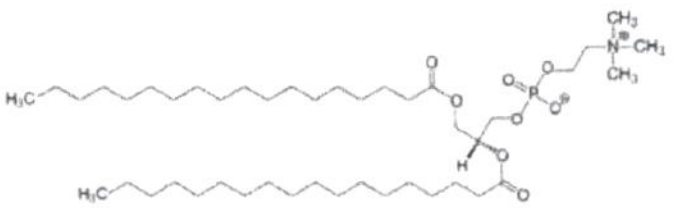

Question 5

The diagrams below illustrate three different levels of protein structure related to folding and assembly. The diagrams are not drawn to scale.

a)   Name the structural levels of protein shown in the diagrams. (3)

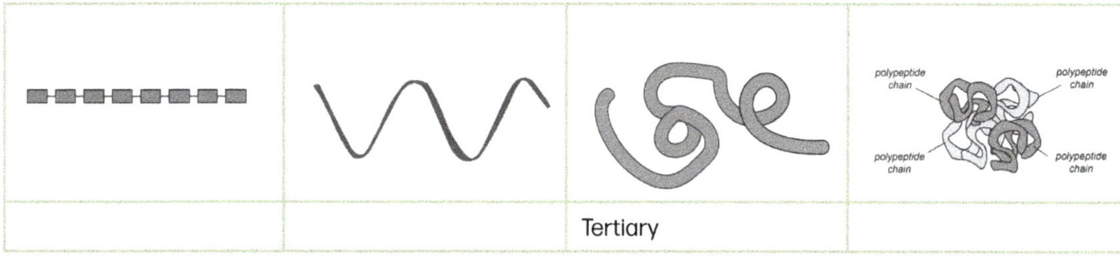

b)   Name the monomer subunit of a protein. (1)

____

Question 6

Describe a polysaccharide found in animals and state its function. (2)

____

____

____

Question 7

Explain why triglycerides are not strictly regarded as polymers. (1)

____

____

Question 8

Describe the key structural differences between triglyceride and phospholipid molecules. (2)

____

____

____

____

Question 9

Differentiate between saturated and unsaturated fatty acids. (1)

____

____

Question 10

Complete the table below by indicating with a tick (✓) which elements are present in carbohydrates, proteins and lipids.

Element	Carbohydrates	Proteins	Lipids
Carbon (C)			
Hydrogen (H)			
Oxygen (O)			
Nitrogen (N)			
Phosphorus (P)			
Sulfur (S)			

Question 11

Energy in living organisms can be stored as carbohydrates or lipids. Name the carbohydrate molecules that are used for energy storage in plants and animals. (2)

Animal ____

Plant ____

Question 12

Describe the tertiary structure of a protein. (3)

____

____

____

Section 2: Enzymes

I can....	Confidently	Somewhat	Not quite
... describe the structure and function of enzymes, including the role of the active site			
...compare the induced-fit and lock-and-key models of enzyme function			
...explain how enzyme activity is affected by factors such as temperature, pH, presence of inhibitors and substrate concentration			
...explain how metabolic processes, such as digestion, are controlled and regulated by enzymes			
...describe the roles of amylase, protease and lipase in chemical digestion			
...interpret data from an experiment investigating factors affecting enzyme activity			

Question 1

The figure shows an enzyme, its substrate and a non-competitive inhibitor.

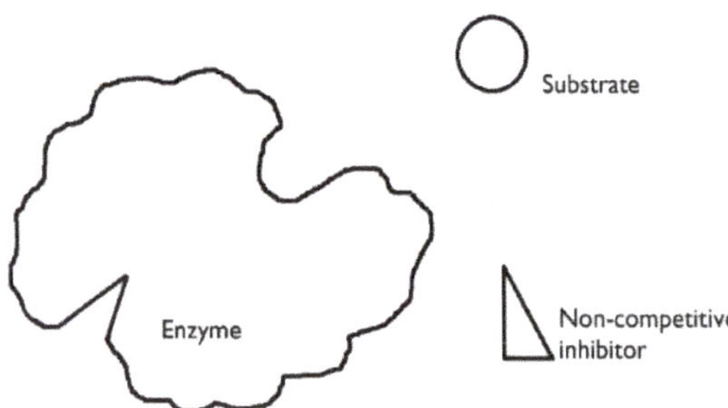

With reference to the figure:

a)   Label the active site on the figure. (1)

b)   Explain how the inhibitor influences the action of the enzyme. (3)

____

____

____

Question 2

Enzymes are biological catalysts. Define the term catalyst. (2)

____

____

Question 3

Enzymes are proteins with a tertiary structure. Explain the importance of the tertiary structure to the functioning of the enzyme. (4)

____

____

____

____

Question 4

Explain how a significant increase in temperature above the optimum affects the rate of an enzyme-controlled reaction. (4)

____

____

____

____

Question 5

During digestion, macromolecules are broken down into their constituent monomers. Complete the table below. (3)

Macromolecule	Enzyme involved	Monomers
		Maltose
	Protease	
Lipids		

Question 6

Living organisms cannot survive without enzymes.

a)   Explain why enzymes are essential for the survival of living organisms. (1)

____

____

b)   Describe the 'active site' of an enzyme and explain its function. (2)

____

____

____

Question 7

Explain how competitive inhibitors affect the activity of an enzyme. (2)

____

____

____

____

Question 8

A fixed volume and concentration of both substrate and enzyme were mixed, while all other variables were kept constant. The enzyme-catalysed reaction was allowed to proceed to completion. Identify the graph that correctly shows how the rate of reaction changes over time. (1)

a) 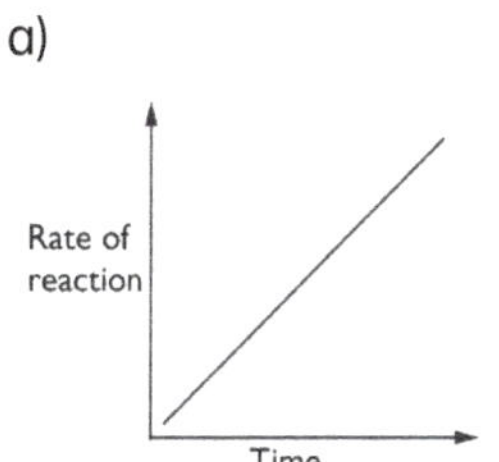

b) 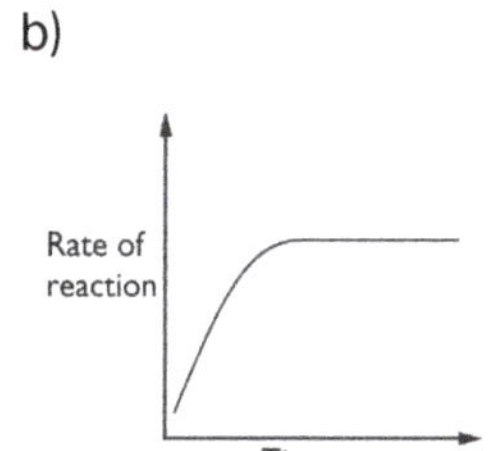

c) 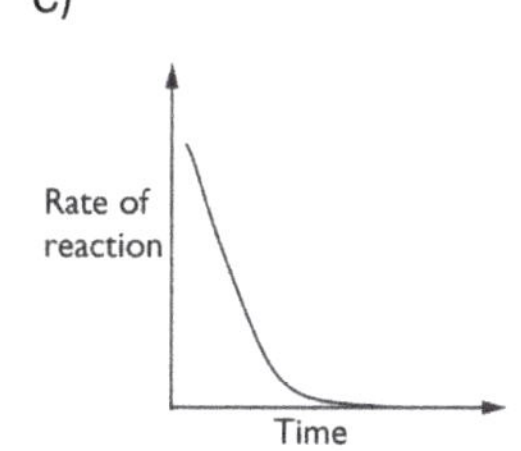

d) 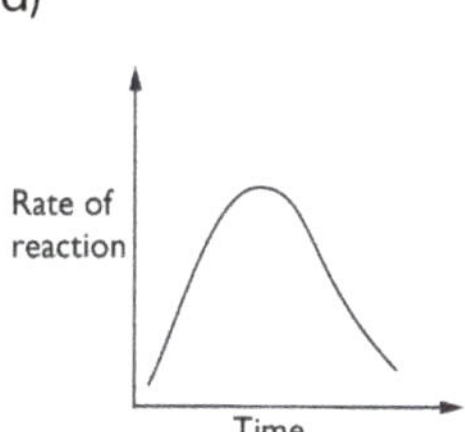

Question 9

Yeast cells produce the enzyme invertase, which catalyses the breakdown of sucrose into glucose and fructose. Since enzymes are proteins, their function is closely related to their structure.

a)   Explain how the structure of an invertase enzyme is related to its ability to catalyse the breakdown of sucrose. (2)

____

____

____

____

The graph shows the activity of invertase at different pH values.

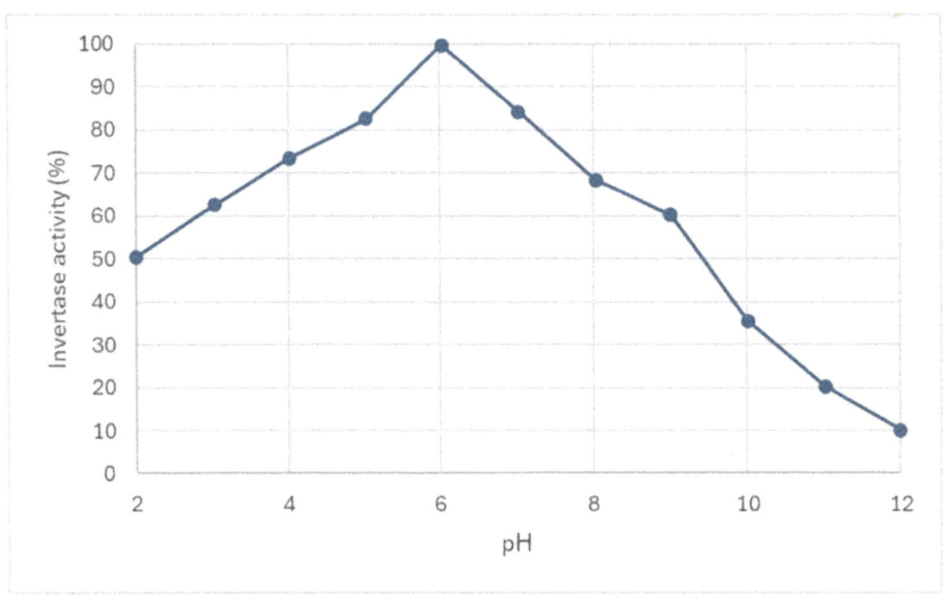

b) Describe and explain the effect of pH on the activity of invertase. (5)

____

____

____

____

____

____

Question 10

Compare the lock and key and the induced fit model. You may include diagrams in your answer. (2)

____

____

____

____

____

____

Question 11

Pectinase is an enzyme that breaks down pectin. A study investigated how various pH and temperature combinations affect pectinase activity. The volume and concentration of both pectinase and pectin were kept constant across all reactions. The table presents the reaction rate for each combination of pH and temperature tested.

pH	Temperature (oC)	Rate of reaction (a.u)
2	10	0
2	35	0
2	70	0
4	10	10
4	35	45
4	70	0
8	10	10
8	35	20
8	70	0

Draw two conclusions from these results. (2)

____

____

____

Question 12

Explain how changes in temperature can affect enzyme action. (4)

____

____

____

____

Section 3: Exchange Surfaces

I can....	Confidently	Somewhat	Not quite
...explain how structural features of exchange surfaces in the digestive and circulatory systems of mammals (e.g. villi and capillaries) allow for efficient nutrient exchange			
...describe how closed circulatory systems facilitate the efficient transport of materials to and from all cells in the body			

Question 1

State the function of villi. (1)

____

____

Question 2

Explain how the structure of villi in the small intestine facilitates efficient nutrient absorption. (2)

____

____

____

Question 3

In a closed circulatory system, blood remains contained within blood vessels. State two advantages of keeping blood inside vessels. (2)

____

____

____

Question 4

Explain how the structure of capillaries is related to their function. (2)

____

____

____

____

Question 5

The figure shows a villus, found in the small intestine.

a) Name X, Y and Z. (3)

X ______________________________________

Y ______________________________________

Z ______________________________________

b) The epithelial cells of the villi absorb nutrients through both diffusion and active transport.
Explain how active transport differs from diffusion. (3)

____

____

____

c)   Explain the significance of part X in nutrient exchange. (2)

____

____

Question 6

Describe what is meant by a closed circulatory system. (2)

____

____

____

Question 7

Coeliac disease triggers an immune response to gluten, a protein in wheat, leading to blunted or atrophic villi in the small intestine. This condition involves the shrinking and flattening of the villi due to repeated gluten exposure, as shown in the figure.

Explain why children with untreated coeliac disease may experience slower growth and increased fatigue compared to those without the condition. Use information from the diagram and your own knowledge to support your answer. (4)

____

____

____

____

Question 8

The figure shows some cells taken from the lining of the small intestine.

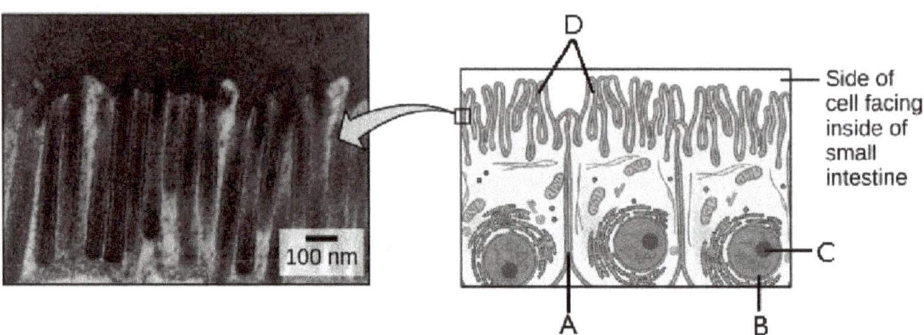

a)   Identify a feature that indicates the cells are from animal tissue. (1)

____

b)   Name the labelled parts of the cells. (4)

A   ____

B   ____

C   ____

D   ____

Section 4: Excretory System

I can....	Confidently	Somewhat	Not quite
...identify the parts of a nephron and their functions in the production of urine			
...explain how glomerular filtration, selective reabsorption and secretion across nephron membranes contribute to the removal of waste			

Question 1

The figure shows the functional unit of a mammalian kidney.

a) Name the functional unit. (1)

____

b) Name S to Z. (7)

S _______________________________________

T _______________________________________

U _______________________________________

V _______________________________________

W _______________________________________

X _______________________________________

Y _______________________________________

Z _______________________________________

c)   State the function of part Y. (1)

____

d)   Mark on the figure where selective reabsorption occurs. (1)

e)   Predict how part Y would differ in mammals that live in the desert. Give a reason for your answer. (2)

____

____

____

f)   Describe how glomerular filtration (ultrafiltration) occurs in a kidney. (3)

____

____

____

Question 2

Explain how the nephron and collecting duct contribute to the higher concentration of urea in urine compared to glomerular filtrate. (5)

____

____

____

____

____

Question 3

The figure shows part of a nephron.

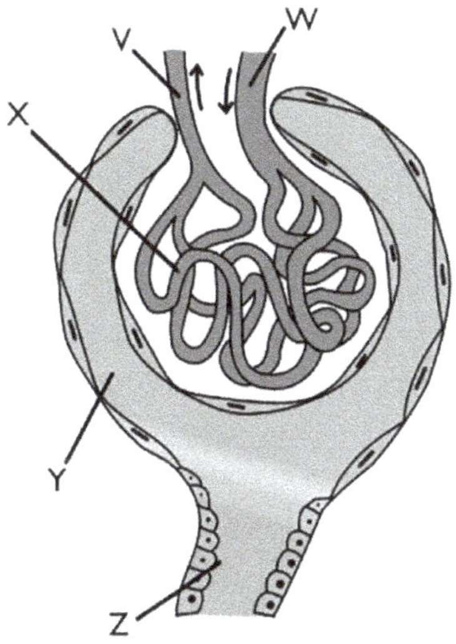

a) Name V to Z. (5)

V _______________________________

W _______________________________

X _______________________________

Y _______________________________

Z _______________________________

b) Other than urea and salts, state two other substances that would be found at Z. (2)

____

c) Name the process that can separate these substances from the rest of the blood plasma. (1)

____

d) Discuss the importance of the difference in diameter between structure V and structure W. (2)

____

____

Question 4

If kidney damage is suspected, testing the patient's urine for the protein albumin is common.
Explain why the presence of albumin in the urine suggests kidney damage. (2)

____

____

TOPIC 3: CELLULAR ENERGY, GAS EXCHANGE AND PLANT PHYSIOLOGY

Section 1: Respiration

I can....	Confidently	Somewhat	Not quite
...distinguish between catabolism and anabolism			
...explain how ATP allows energy from catabolic reactions to be used in anabolic reactions			
...describe the process of aerobic respiration, identifying the location in the cell and net inputs and outputs of each stage			
...compare aerobic and anaerobic respiration			

Question 1

Write a balanced symbol equation for respiration. (1)

____

Question 2

Distinguish between catabolism and anabolism. (2)

____

____

____

Question 3

Give an example of where ATP is used in the following types of reaction. (2)

Catabolic reaction ____

Anabolic reaction ____

Question 4

State an advantage of aerobic respiration over anaerobic respiration. (1)

____

____

Question 5

Yeast is a single-celled organism capable of aerobic respiration, which occurs in the mitochondria of yeast cells. Name two molecules required for aerobic respiration that can enter the mitochondria. (2)

____

____

Question 6

In aerobic respiration, glucose is broken down in three main stages, as outlined in the figure.

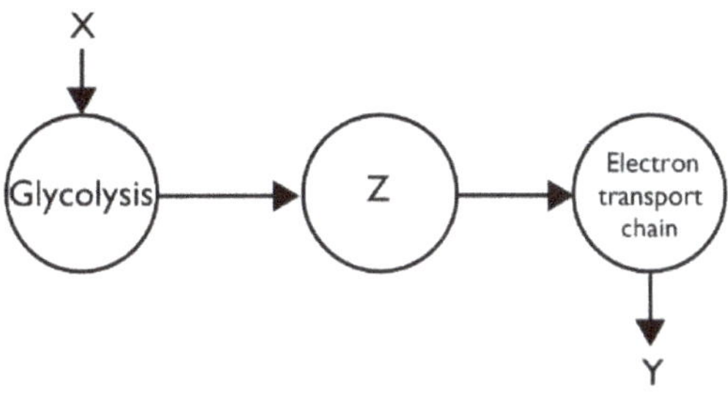

a)   Name molecules X and Y, and stage Z. (3)

X   ____

Y   ____

Z   ____

b)   State where the electron transport chain takes place. (1)

____

Question 7

Complete the following table to compare aerobic and anaerobic respiration. (7)

Feature	Aerobic Respiration	Anaerobic Respiration
Oxygen requirement		
Location in cells		
Energy yield	High	
End products		
Efficiency		
ATP production		2 per glucose molecule
Process stages	Glycolysis, Krebs cycle, Electron Transport Chain	

Question 8

Explain how ATP allows energy from catabolic reactions to be used in anabolic reactions. (2)

____

____

____

Question 9

Identify the location in the cell of each stage of respiration, in addition to the net inputs and outputs of each stage. (6)

	Glycolysis	Krebs Cycle	Electron transport chain
Location in cell			
Net inputs			
Net outputs			

Question 10

Yeast is a single-celled microorganism that grows in a nutrient-rich liquid inside a fermenter. It can perform both aerobic and anaerobic respiration.

a) State the word equation for aerobic respiration. (1)

____

b) State the word equation for anaerobic respiration in yeast. (1)

____

c) Describe how anaerobic respiration in animal cells is different to that in yeast. (1)

____

____

d) Name an example of when animals would need to respire anaerobically. (1)

____

____

Section 2: Photosynthesis

I can....	Confidently	Somewhat	Not quite
...describe the process of photosynthesis, identifying the location in the cell and net inputs and outputs of the light-dependent and light-independent reactions			

Question 1

a) Write a balanced symbol equation for photosynthesis. (1)

____

b) Name two products of the light-dependent reactions of photosynthesis that are utilised in the light-independent reactions. (2)

____

____

c) The diagram shows a chloroplast.

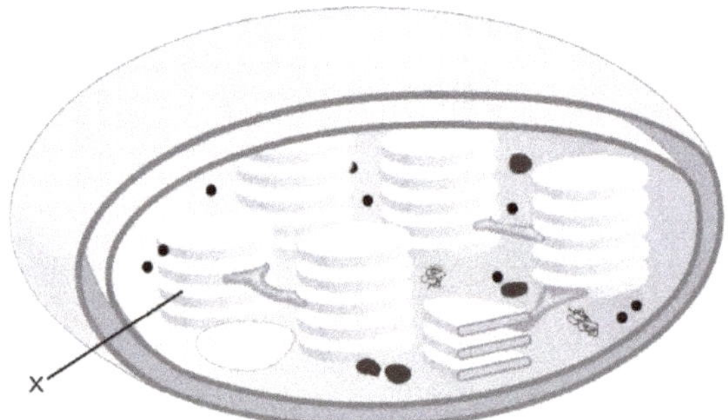

i) Name structure X. (1)

____

ii) Complete the following table. (3)

Name of the stage of photosynthesis occurring at X			
Input molecules required at X	1.		2.
Output molecules produced at X	1.		2.

Question 2

Scientists are exploring methods to enhance the efficiency of photosynthesis in plants, particularly in how carbon dioxide is captured.

a) Identify the stage of photosynthesis where carbon dioxide is captured. (1)

____

b) This stage requires additional inputs. Name two of these inputs and explain the role of each in this stage of photosynthesis. (2)

Input	Role

c) Explain the role of the electron transport chain in photosynthesis. (2)

____

____

d) Identify the location in the cell of the light-dependent and light-independent stages of photosynthesis, in addition to the net inputs and outputs of each stage. (6)

	Light-dependent stage	Light-independent stage
Location in cell		
Net inputs		
Net outputs		

Section 3: Gas Exchange

I can....	Confidently	Somewhat	Not quite
...explain how structural features of exchange surfaces in the respiratory and circulatory systems of mammals (alveoli and capillaries) allow for efficient gas exchange			
...analyse data to predict the direction that materials will be exchanged between alveoli and capillaries; and capillaries and muscle tissue			

Question 1

The gaseous exchange surface of lungs possesses several features which allow efficient gas exchange. State how each of the following features allows for efficient gaseous exchange.

a) The surface is composed of many alveoli. (1)

____

____

b) Alveolar walls are 0.1μm thick. (1)

____

____

c) A network of capillaries covers the outside of each alveolus. (1)

____

____

Question 2

Partial pressure is the pressure exerted by a single gas in a mixture. It drives the movement of gases from areas of high partial pressure to low partial pressure through diffusion. This principle is key in understanding how oxygen and carbon dioxide are exchanged in the lungs and tissues, allowing oxygen to enter the blood and carbon dioxide to be removed.

The partial pressure of oxygen in the alveoli is 14 kPa and in the pulmonary capillaries it is 6 kPa. Describe how this will affect the direction of the diffusion of oxygen. (1)

____

____

Question 3

The figure shows an alveolus with associated blood capillaries.

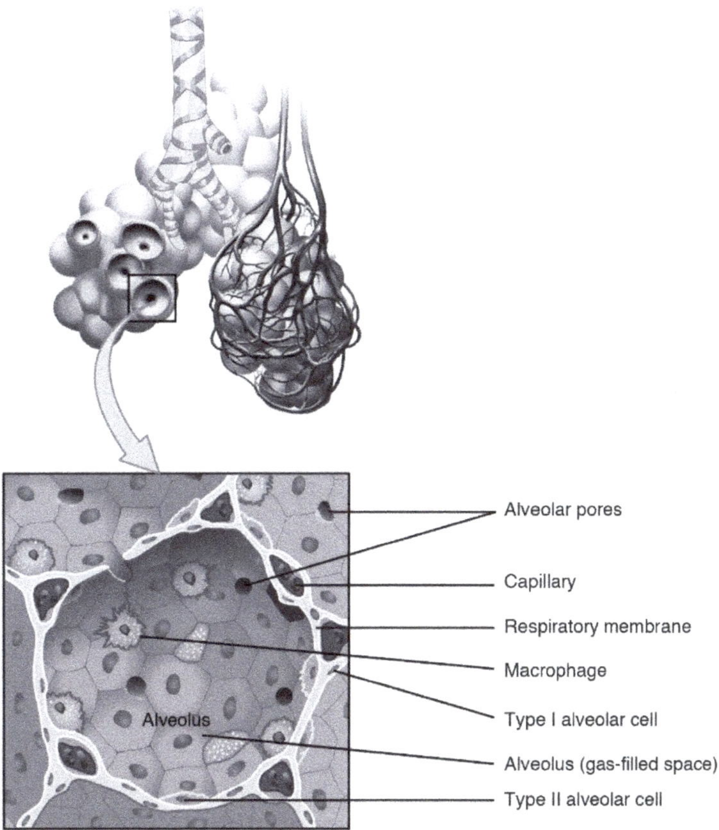

a) State a feature, visible in the figure, that shows that alveolar cells are eukaryotic. (1)

____

b) There are two types of alveolar cells: type I and type II. Type I alveolar cells are the squamous cells responsible for gas exchange between the alveoli and blood. Type II alveolar cells secrete surfactant proteins. Propose what organelle/s would be found in higher amounts in Type II cells than in Type I cells. Give a reason for your answer. (2)

____

____

c) State two features of a gas exchange surface, such as the lining of the alveolus. (2)

____

____

d) When oxygen passes from the alveolus to red blood cells, it diffuses through plasma membranes. Describe how other molecules or ions cross a plasma membrane by active transport and facilitated diffusion. (6)

____

____

____

____

e) Red blood cells have a biconcave shape. Explain one advantage of this shape over a spherical cell that has the same volume. (2)

____

____

____

Section 4: Plant Physiology

I can....	Confidently	Somewhat	Not quite
...compare the structure and function of xylem and phloem tissues			
...explain how stomata and guard cells facilitate gas exchange in plant			
...interpret data from an experiment investigating the effect of light intensity, temperature, wind or humidity on the rate of transpiration			

Question 1

Define the term transpiration. (2)

____

____

Question 2

Explain why xylem is described as a tissue. (2)

____

____

Question 3

Complete the table to compare the structure and function of xylem and phloem tissues. (4)

	Xylem	Phloem
Function	Transports water and dissolved minerals from roots to other parts of the plant.	
Cell types	Includes vessel elements, tracheids, xylem parenchyma and xylem fibres.	

Direction of flow		Bidirectional
Living/dead cells	Contains both living (xylem parenchyma) and dead cells (vessels and tracheids).	

Question 4

Complete the table to show how different environmental factors affect the rate of transpiration. (4)

Condition	Effect on transpiration rate	Explanation
Increased temperature		
Increased wind speed		
Increased humidity		
Decreased light intensity		

Question 5

Stomata on leaf surfaces are the primary sites for gas exchange in plants. Describe the mechanism of stomatal opening that facilitates this gas exchange. (3)

____

____

____

____

Question 6

The table shows the results from some experiments considering how light intensity and temperature can impact the rate of transpiration in plants.

Condition	Light Intensity (lux)	Temperature (°C)	Transpiration Rate (g/hr)
Control	1000	20	0.3
Low Light	500	20	0.15
High Light	2000	20	0.5
Low Temperature	1000	10	0.1
High Temperature	1000	30	0.55

Analyse the data to determine how light intensity and temperature affect the rate of transpiration. (3)

____

____

____

____

____

____

____

____

Unit 2

Maintaining the internal environment

TOPIC 1: HOMEOSTASIS - THERMOREGULATION AND OSMOREGULATION

Section 1: Nervous System

I can....	Confidently	Somewhat	Not quite
...explain how the nervous system uses negative feedback to coordinate responses to internal/external stimuli and maintain homeostasis (stimulus-response model)			
...identify the different types of sensory receptors and their stimuli			
...describe the structure and function of nerve cells			
...distinguish between sensory neurons, interneurons and motor neurons			
...explain the passage of a nerve impulse in terms of transmission of an action potential			
...explain the passage of a nerve impulse in terms of synaptic transmission			

Question 1

Describe the process by which a nerve impulse arriving at a synapse leads to the release of neurotransmitters from vesicles in the presynaptic membrane. (4)

____

____

____

Question 2

Complete the passage. (3)

Specialised cells that detect ____________________ are located both internally and externally. These cells are referred to as sensory ____________________. Each is tailored to respond to a specific type of ____________________.

Question 3

Explain how a resting potential of -70 mV is maintained in a neuron. (2)

____

____

____

Question 4

The diagram shows cells involved in detecting and responding to a stimulus.

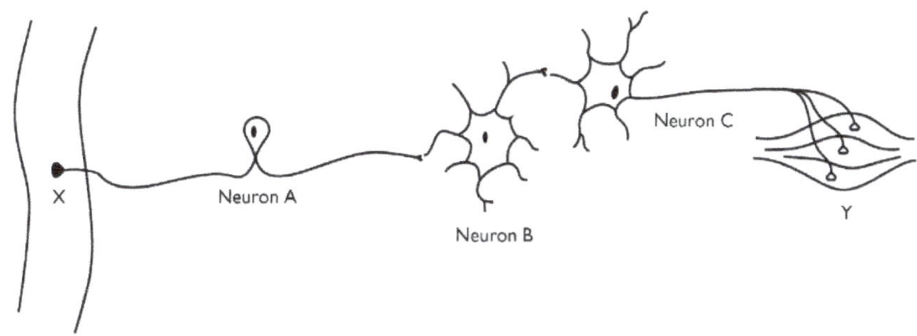

a)   Name the parts of the diagram. (5)

Neuron A   ____

Neuron B   ____

Neuron C   ____

X   ____

Y   ____

b)   Using an arrow, mark the position of a synapse on the diagram. (1)

Question 5

The diagram shows a neuron.

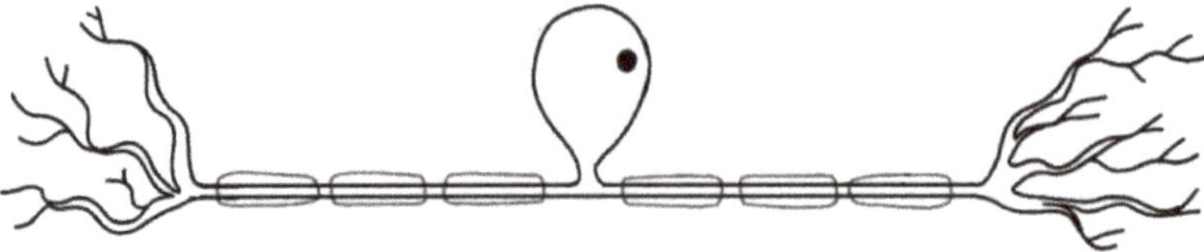

a)   Identify the type of neuron shown in the diagram. (1)

____

b)   How does the position of the soma (cell body) in other types of neurons compare to the position illustrated in the diagram? (1)

____

Question 6

State the name given to the gap between the two neurons. (1)

____

Question 7

Involuntary actions occur through nerve impulses that travel along reflex arcs. An example is the quick withdrawal of a hand after pricking yourself on a pin. The figure illustrates the structures involved in this reflexive hand movement.

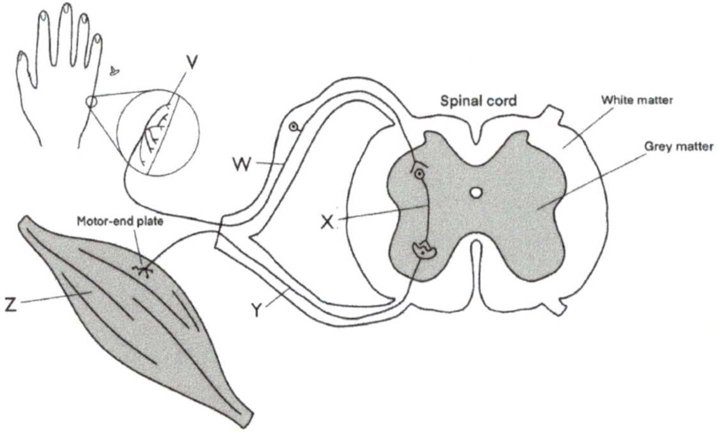

The table lists the functions of certain structures shown, along with their names and corresponding letters from the figure. Complete the table with the missing information.

Name	Letter	Function
	V	
	W	
	X	
	Y	
	Z	

Question 8

State the term that describes maintaining a constant internal environment by negative feedback. (1)

____

Question 9

The graph illustrates the changes in membrane potential across a neuron's membrane during the transmission of an action potential.

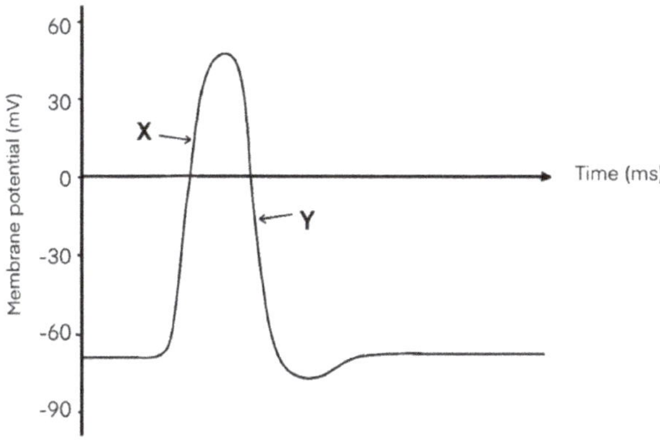

a)   Identify the resting potential from the graph. (1)

____

b) Name the term used to describe what is happening at X and Y. (2)

X ____

Y ____

c) Describe what is happening at X and Y. (4)

X ____

   ____

   ____

Y ____

   ____

   ____

Question 10

Complete the passage below by filling in the blanks with the most appropriate terms. (3)

When an impulse is not passing along a neuron, a resting potential of _______________ mV
is established. When the neuron is stimulated, it causes _______________ of the cell surface
membrane. This will not generate an action potential unless it is large enough to exceed the _
_______________________________.

Question 11

The nervous system consists of various types of neurones that transmit electrical impulses.
Complete the table below by listing three structural differences between motor and sensory
neurons. (3)

Sensory neurons	Motor neurons

Question 12

Complete the table to show the different types of sensory receptors and their stimuli. (6)

Sensory receptor	Stimulus detected
Photoreceptor	
	Mechanical pressure, vibration, touch
Thermoreceptor	
	Pain
Chemoreceptor	
	Pressure changes

Question 13

Tetrodotoxin is a neurotoxin found in pufferfish. It blocks voltage-gated sodium channels in neurons. Explain how the presence of tetrodotoxin at the synapse affects the initiation and propagation of action potentials in neurons. (3)

Question 14

Explain why a myelinated axon can conducts impulses faster than a non-myelinated axon of the same diameter. (3)

Section 2: Endocrine System

I can....	Confidently	Somewhat	Not quite
...explain how the endocrine system uses negative feedback to coordinate responses to internal/external stimuli and maintain homeostasis (stimulus-response model)			
...describe how hormones relay messages to cells displaying specific receptors via the circulatory or lymphatic system			
...explain how receptor binding alters cellular activity, recognising that a cell's sensitivity to a specific hormone is directly related to the number of receptors it displays for that hormone			
...analyse feedback-control diagrams to identify the stimulus, receptor/s, control centre, effector/s and communication pathway/s in different scenarios			

Question 1

Define homeostasis. (1)

____

Question 2

A hormone was produced in one cell, entered the bloodstream and travelled to two adjacent groups of cells. One group responded to the hormone while the neighbouring group did not. Explain the reason for the difference in response to the same hormone between these cells. (1)

____

____

____

____

Question 3

Negative feedback mechanisms are crucial for maintaining homeostasis. Describe the key features of a negative feedback mechanism. (2)

____

____

____

Question 4

Explain how the human body responds to high blood glucose levels. (2)

____

____

____

Question 5

The stimulus-response model is made up of several stages, represented by the boxes in the figure below. Complete the diagram by adding correct words to the boxes. One stage is already labelled. (4)

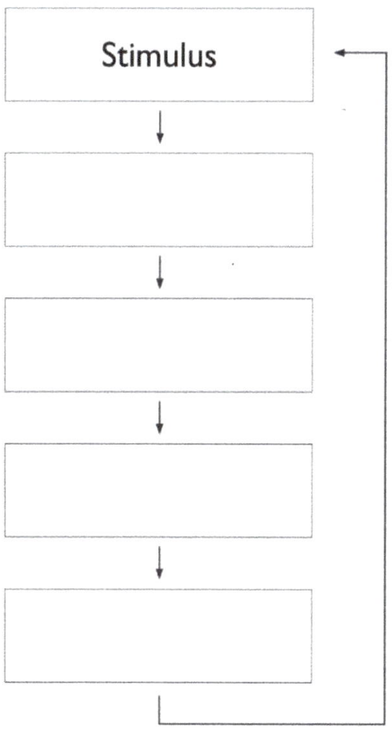

Question 6

There are various types of endocrine cells in complex multicellular animals, each secreting a unique chemical.

a) Explain how each type of endocrine cell affects only specific cell types within the body. (1)

____

____

b) A tissue may become less responsive to a particular hormone over time, even if the hormone concentration remains the same. Suggest a reason for this decrease in responsiveness. (1)

____

____

Question 7

Scientists conducted a study on how blood glucose levels are regulated in rats. They deprived a group of normal rats of food for 24 hours. Despite this prolonged fasting period, the blood glucose levels in the rats remained stable, similar to their levels at the beginning of the study. Explain how the normal rats maintained their blood glucose levels, despite not eating for 24 hours. (3)

____

____

____

____

____

____

Question 8

When exercising, the body's demand for oxygen increases, causing the heart rate to rise. The baroreceptors in the aorta and carotid arteries sense this change and send signals to the medulla oblongata in the brain. The medulla adjusts the heart rate by sending signals to the heart. Draw a feedback-control diagram to identify the stimulus, receptor/s, control centre, effector/s and communication pathway/s in this scenario. (2)

Section 3: Thermoregulation

I can....	Confidently	Somewhat	Not quite
...explain thermoregulatory mechanisms of endotherms			
...explain thermoregulation in humans using feedback control diagrams			

Question 1

The nervous and endocrine systems work together to keep the body's internal environment within tolerance limits. One key example of this is thermoregulation, which involves maintaining the body's internal temperature.

Explain how the nervous and endocrine systems work together to regulate the body's internal temperature when it risks rising above the tolerance limits. Include two nervous mechanisms and one endocrine mechanism in your answer. (10)

____

____

____

____

____

____

____

Question 2

Define the term endotherm. (1)

____

Question 3

Many organisms use evaporative cooling for thermoregulation. Explain the key principles of evaporative cooling. (4)

____

____

____

____

Question 4

The polar bear inhabits the Arctic tundra, one of the coldest environments on Earth. Discuss one structural feature and one physiological process that help mammals living in cold environments maintain a constant core body temperature. (4)

____

____

____

____

Question 5

Describe how adjusting the volume of blood flowing through the skin assists in regulating body temperature. (4)

____

____

____

Question 6

Correctly match the endothermic thermoregulatory mechanisms with the correct definition. (4)

Aestivation	A state of reduced metabolic rate that allows animals to conserve energy during periods of reduced food availability or adverse conditions.
Torpor	Where an animal relies on the body heat of another animal or its environment to regulate its own body temperature.
Kleptothermy	A period of dormancy or inactivity that some animals enter during hot or dry conditions to conserve energy and avoid adverse environmental conditions.
Hibernation	A prolonged state of inactivity and reduced metabolic rate that animals enter during cold periods to conserve energy and survive through winter when food is scarce.

Question 7

Adipose tissue in mammals exists in two types: white adipose tissue and brown adipose tissue, with their distribution varying among species. For example, polar bears, which live in icy environments, have a thick layer of white adipose tissue beneath their skin. Additionally, other parts of a polar bear's body contain a high proportion of brown adipose tissue. The structure of each type of fat cell in polar bears is adapted to help them survive in cold conditions.

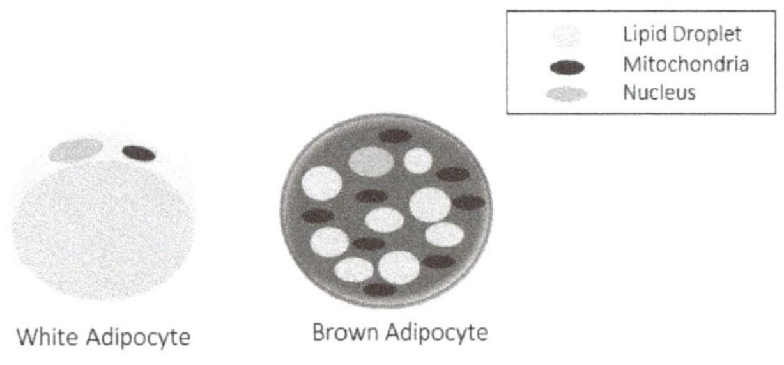

Using the information in the figure, explain how the structure of white and brown fat cells supports the survival of polar bears in their cold habitat. (2)

____

____

____

Section 4: Osmoregulation

I can....	Confidently	Somewhat	Not quite
...explain osmoregulation in humans, including the role antidiuretic hormone (ADH) and the kidney using feedback control diagrams			
...explain how structural and homeostatic mechanisms maintain water balance in plants			
...interpret data from an experiment comparing the number and distribution of stomata in plants adapted to different environments			

Question 1

Name the cells that control the opening and closing of stomata. (1)

____

Question 2

ABA is known to promote the closure of stomata. When ABA binds to receptors on the surface of guard cells, it triggers the removal of ions from these cells. Explain how the loss of ions in guard cells leads to the closure of stomata. (2)

____

____

____

Question 3

Plants lose water through their leaves.

a) Name the cells in a leaf that regulate the rate of water loss. (1)

____

b) Water is absorbed by the roots, transported through the plant and evaporated from the leaves. State the term that describes this movement of water. (1)

____

c)   Name a change that would reduce the rate of water loss from a plant's leaves. (1)

____

Question 4

The figure shows a cross section of a leaf of a xerophyte.

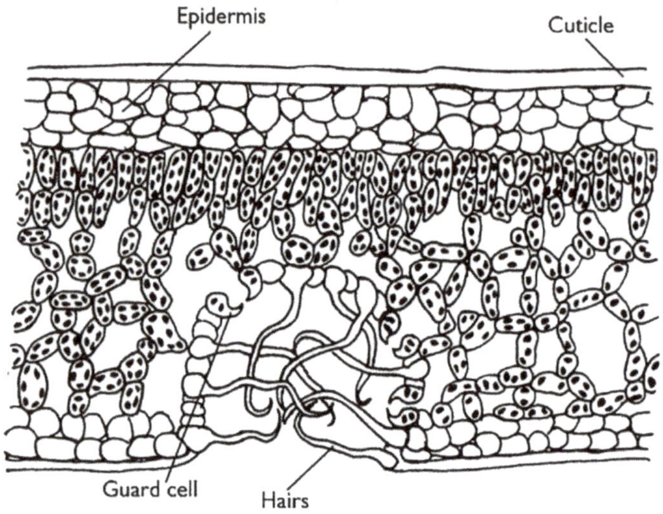

Epidermis

Cuticle

Guard cell

Hairs

a)   Explain how the cuticle reduces water loss. (1)

____

____

b)   Explain how the guard cells and hairs can also reduce water loss. (4)

____

____

____

____

____

Question 5

Osmoreceptors are specialised cells that detect changes in the blood's water potential.

a)   Identify their location in a mammal's body. (1)

____

Stimulation of osmoreceptors can lead to secretion of ADH.

b)   Name the gland that releases ADH. (1)

____

c)   Explain how the secretion of ADH affects urine production. (3)

____

____

____

Question 6

Many organisms live in arid environments.

a)   State the name of plants that are adapted to arid environments? (1)

____

b)   Describe how plants lose water to the environment. (2)

____

____

____

Question 7

The kidneys regulate the water potential of body fluids through a process called osmoregulation, which relies on a negative feedback mechanism. Describe how negative feedback functions in osmoregulation. (4)

____

____

____

____

____

Question 8

The width of stomatal openings in plants is regulated by the movement of ions. These ions move through channel proteins in the cell surface membranes of the guard cells.

a) Draw a diagram showing part of a cell surface membrane with a channel protein. Label your diagram. (3)

b) Explain why channel proteins are required for the movement of ions into and out of cells. (2)

____

____

____

c) The outward movement of ions from guard cells causes stomata to close. Some plants have
    been genetically engineered to lack channel proteins for the outward movement of ions in the
    cell surface membranes of the guard cells. Explain the effect of this absence on transpiration. (3)

____

____

____

Question 9

The data shows the mean number of stomata per mm^2 in the upper and lower epidermis of a
water lily and a eucalypt.

Plant	Mean number of stomata per mm2	
	Upper epidermis	Lower epidermis
Water lily	350	0
Eucalypt	20	170

Explain the differences between the mean number of stomata found in a water lily and a
eucalypt. (5)

____

____

____

____

____

TOPIC 2: INFECTIOUS DISEASE AND EPIDEMIOLOGY

Section 1: Disease

I can....	Confidently	Somewhat	Not quite
...distinguish between infectious and non-infectious disease			
...identify key features of prions, viruses, bacteria, fungi, protists and parasites			
...explain how adherence factors, invasion factors, capsules and toxins affect pathogenesis			

Question 1

Explain how infectious diseases differ from other types of diseases. (2)

____

____

Question 2

Complete the table below by identifying the classification of the following types of pathogens as either cellular or non-cellular, and if cellular, specify if they are eukaryotic or prokaryotic. (5)

Pathogen	Cellular or non-cellular	Eukaryotic or prokaryotic
Prion		
Virus		
Bacterium		
Fungus		
Protist		

Question 3

Describe one adaptation utilised by pathogens that helps them to enter a host. (2)

____

____

____

Question 4

State how endotoxins and exotoxins differ. (1)

____

____

Question 5

Define the term pathogen. (1)

____

____

____

Question 6

Name the following pathogens. (4)

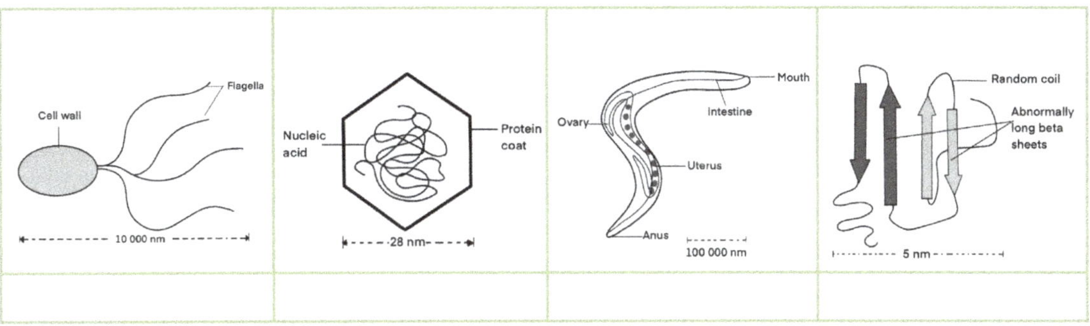

Question 7

Streptococcus pneumoniae is a bacterium that can cause severe infections such as pneumonia, meningitis and sepsis. It possesses both a thick slime capsule and pili.

Describe how each of these structures aid in pathogenesis. (2)

____

____

____

____

Question 8

The pathogenesis of anaerobic infections involves the disruption of the mucosal surface, allowing anaerobic bacteria to enter and invade deeper tissues. The site and severity of the infection are influenced by both the virulence factors of the bacteria and the host's immune response.

Use lines to correctly match the virulence factor that aids pathogenesis with the correct example. (3)

Adherence factor	Collagenase and hyaluronidase
Invasion factor	Botulinum neurotoxins (BoNTs)
Toxin	Pili and fimbriae

Section 2: Immune Response

I can....	Confidently	Somewhat	Not quite
...explain how host cells recognise self from non-self			
...identify the three lines of defence in vertebrates			
...describe the inflammatory response			
...explain the adaptive immune response			
...compare active and passive immunity, both naturally acquired and artificially acquire			
...interpret long-term immune response data			
...describe the innate immune responses in plants			
...interpret data from an experiment investigating the effect of an antimicrobial agent on the growth of a microorganism			

Question 1

B lymphocytes are distributed throughout the blood and lymphatic system. When an antigen enters the body, some B lymphocytes transform into plasma cells that produce antibodies.

a) Explain the difference between an antigen and an antibody. (4)

____

____

____

b) Explain why antibody production is faster and higher when the same antigen enters the body for a second time. (4)

____

____

____

____

Question 2

People can acquire active and passive immunity to diseases like chickenpox. Describe one active and one passive way in which babies can become immune to chickenpox. (2)

Active ____

Passive ____

Question 3

The table shows some plant defences. Use ticks (✓) to identify if the defence is chemical or physical. (4)

Plant defence	Type of defence	
	Chemical	Physical
Thick, waxy cuticle		
Plant defensins		
Poisonous berries		
Bark that falls off		

Question 4

Researchers testing new vaccines for malaria are conducting trials in animal subjects. To assess the effectiveness of a new vaccine, they measure both humoral and cell-mediated immune responses in the animals. Explain how these two immune responses differ, providing two differences in your answer. (2)

____

____

____

____

Question 5

Name and describe the function of one type of immune cell that contributes to the innate immune response. (2)

____

____

Question 6

Explain how immune system cells distinguish between self and non-self. (1)

____

____

Question 7

Dengue fever is caused by a virus transmitted through the bite of a specific mosquito species.

a)   Are viruses classified as a cellular or non-cellular pathogens? Justify your answer. (2)

____

____

____

b)   An unvaccinated individual travelled to a region where the dengue virus is present and was exposed to the virus. Describe two ways in which their body's first line of defence would act to protect against infection by this virus. (2)

____

____

____

Question 8

Explain how vaccination protects a person from disease. (2)

____

____

____

Question 9

Describe how complement proteins and natural killer cells help protect the human body after a pathogen has overcome the first line of defence. (2)

____

____

____

Question 10

Describe two features of the inflammatory response that help reduce the impact of an infection. (2)

____

____

____

Question 11

Describe an antigen. (1)

____

____

Question 12

The disc diffusion agar method tests antibiotic effectiveness on a microorganism. An agar plate is inoculated with bacteria, and paper discs with antibiotics are added. After bacteria grow, the inhibition zones around each disk indicate effectiveness. The figure shows six antibiotics being tested.

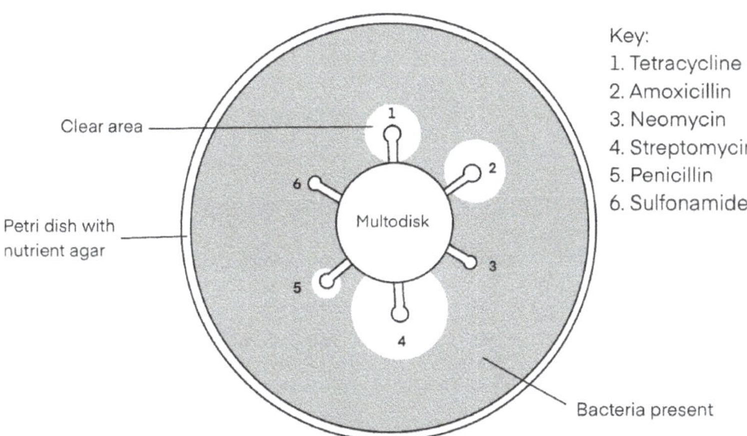

Draw conclusions about the antibiotics used in the experiment, providing evidence from the figure to support your answer. (4)

____

____

____

____

Question 13

The inflammatory response is a non-specific response to infection. Explain how alterations in blood vessels contribute to the redness and swelling observed at the site of inflammation. (4)

____

____

____

____

____

____

Question 14

Non-specific immune responses feature the enzyme lysozyme, which targets and breaks down bacterial cell walls. The figure illustrates with an arrow where lysozyme breaks a bond in peptidoglycan, a key component of these cell walls.

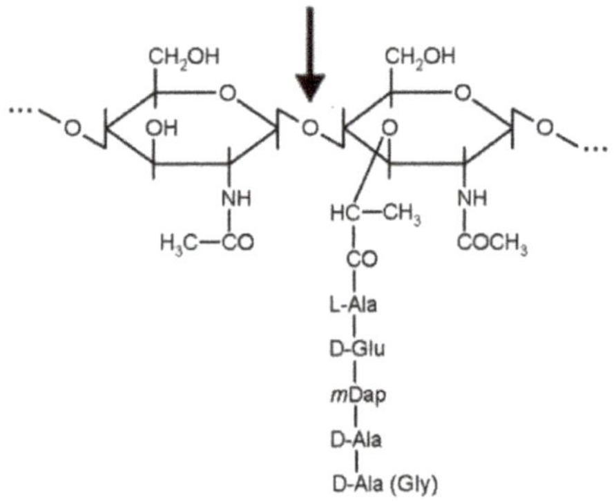

a) Name the bond broken by lysozyme. (1)

____

b) Name another molecule that would be released in this reaction. (1)

____

c) Explain why lysozyme cannot destroy viruses. (2)

____

____

____

Question 15

Describe how phagocytes such as neutrophils and macrophages can destroy pathogens. (1)

____

____

Question 16

When a person is exposed to rabies through an animal bite, a rabies virus enters the body. As an immediate treatment, rabies immunoglobulin is injected. This immunoglobulin contains antibodies specific to the rabies virus. This is an example of passive immunity. Explain how the treatment with rabies immunoglobin works and why it is essential to use passive immunity, rather than active immunity. (2)

____

____

____

Question 17

State the name of an innate immune cell that directly destroys virally infected or cancerous cells. (1)

____

Question 18

Statements A to F below outline the stages of an immune response to a virus, listed in random order:

A	T helper cells become activated and undergo mitosis
B	B cells are activated, undergo mitosis and then differentiate
C	Viral antigens are displayed
D	The virus is engulfed
E	Plasma cells produce and release antibodies specific to the antigens
F	T helper cells secrete cytokines

Place the letters A to F representing the statements into the correct order. (2)

________ → ________ → F → ________ → ________ → ________

Question 19

Compare the functions of B and T lymphocytes in the specific immune response. (6)

____

____

____

____

____

____

Question 20

Gene therapy can be used to treat certain genetic disorders by introducing a normal allele into target cells using a vector. In a trial for treating muscular dystrophy, a viral vector was used for gene therapy. This viral vector triggered a primary immune response, resulting in the production of memory cells. Explain why the formation of memory cells might hinder the effectiveness of gene therapy for long-term chronic conditions like muscular dystrophy. (3)

____

____

____

____

____

Question 21

The graph shows the concentration of antibodies in the blood of a person who has been infected with a pathogen.

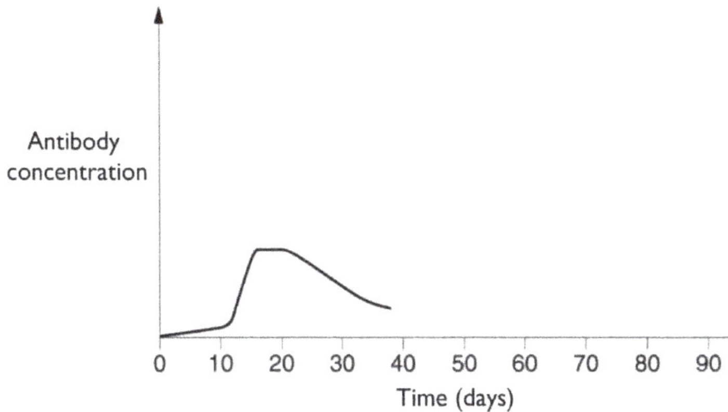

a) The person is exposed to the pathogen again at day 50. Complete the graph. (1)

b) Explain the shape of the graph you have drawn. (3)

____

____

____

____

____

____

____

Section 3: Epidemiology

I can....	Confidently	Somewhat	Not quite
...describe modes of disease transmission			
...explain how different factors affect the spread of disease			
...explain how personal hygiene measures, contact tracing and quarantine are used to control the spread of disease			
... analyse data to predict outbreaks; determine the source of an outbreak; infer the mode of disease transmission; and determine the effectiveness of different strategies in controlling the spread of disease			

Question 1

Infectious diseases occur when pathogens invade body tissues, and these pathogens can be transmitted through various means. The body's external defence mechanisms are the first line of defence against them.

a)   Describe four ways in which pathogens can be transmitted. (4)

____

____

____

____

____

b)   Describe how the body's first line of defence protects against pathogens. (2)

____

____

____

Question 2

Respiratory syncytial virus (RSV) can lead to severe respiratory illness. Name two measures that a person infected with RSV could take to minimise the spread of the virus to others. (2)

____

____

____

Question 3

Explain the process of contact tracing and how it can be used to control a disease outbreak. (3)

____

____

____

____

Question 4

Herd immunity is a strategy used to control the spread of certain infectious diseases. Explain the key principles of herd immunity. (4)

____

____

____

____

Question 5

Explain the process of quarantine and how it is used to control an outbreak of an infectious disease. (3)

____

____

____

____

Question 6

Eating food contaminated with *E. coli* bacteria can lead to illness. Two symptoms of *E. coli* infection are vomiting and diarrhoea. Name two measures a person with a mild *E. coli* infection can take to prevent spreading the bacteria to others. (2)

____

____

____

____

Question 7

Describe some ways to maintain hygiene in the home that will minimise the spread of disease. (3)

____

____

____

Question 8

Analyse how the mobility of individuals within an affected population affects the spread of infectious diseases. Consider factors such as travel patterns, population density, and social interactions in your answer. (4)

____

____

____

____

____

____

Question 9

In managing infectious disease outbreaks, evaluating the effectiveness of various control strategies is crucial for minimising the impact on public health. By analysing data from different intervention measures, we can determine how well each strategy works to reduce the number of cases and control the spread of disease.

Use the data below to evaluate the effectiveness of the vaccination, quarantine and public health campaigns in controlling the spread of the disease. How did you measure their success, and what conclusions can be drawn about their impact? (3)

Control strategy:	Vaccination Campaign	Quarantine Measures	Public Health Campaign
People affected	Coverage: 70% of population	Areas quarantined: 3 high-risk zones	80% of population
Cases before campaign	100	120	80
Cases after campaign	30	50	40

____

____

____

Question 10

An outbreak of a disease has been reported in a town over four weeks. The symptoms include fever, cough and fatigue. Health authorities have collected data on the outbreak, shown in the table below.

Week	Number of cases	Location affected	Control measures implemented
1	5	Market, school	None
2	20	Market, school, gym	Public awareness campaign on hygiene
3	50	Market, school, gym, office	Closure of Market; increased sanitation efforts
4	15	School, office	Quarantine of symptomatic individuals; face mask mandate

Additional Information:

- In Week 1, all initial cases reported visiting the Market.
- In Week 2, more cases were linked to individuals who had visited the Gym and the School.
- Environmental tests showed contaminated surfaces in the Market.

a) Based on the data, predict the trend of cases for Week 5. What factors might influence this trend? (2)

____

____

b) Analyse the data to identify the most likely source of the initial outbreak. Justify your answer with evidence. (2)

____

____

c)   From the pattern of spread and locations affected, infer the most probable mode of disease transmission. (1)

____

____

____

d)   Evaluate the effectiveness of the control measures implemented in Week 3 and Week 4. Which measure seems to have the most impact on reducing the number of cases? (2)

____

____

____

e)   Suggest an additional strategy that could further reduce the spread of the disease, explaining why it would be effective based on the transmission mode inferred.

____

____

____

BIBLIOGRAPHY

https://commons.wikimedia.org/wiki/File:Mitochondrion_186.jpg

https://commons.wikimedia.org/wiki/File:0313_Endoplasmic_Reticulum_b_en.png

https://commons.wikimedia.org/wiki/File:Simple_plant_and_animal_cell.svg

https://commons.wikimedia.org/wiki/File:Mitosis_cells_sequence_English.svg

https://commons.wikimedia.org/wiki/File:Xerophyten_-_Blattanatomie.png

https://commons.wikimedia.org/wiki/File:%D8%AE%D9%85%D9%84%D8%A9_%D9%85%D8%B9%D9%88%D9%8A%D8%A9.svg

https://commons.wikimedia.org/wiki/File:Coeliac_path.jpg

https://commons.wikimedia.org/wiki/File:%CE%95%CE%B9%CE%BA%CF%8C%CE%BD%CE%B1_1._%CE%BC%CF%8C%CF%81%CE%B9%CE%BF_%CF%80%CE%B5%CF%80%CF%84%CE%B9%CE%B4%CE%BF%CE%B3%CE%BB%CF%85%CE%BA%CE%AC%CE%BD%CE%B7%CF%82.jpg

https://commons.wikimedia.org/wiki/File:Figure_1-_Pathway.png

https://commons.wikimedia.org/wiki/File:Sickle_Cell_Anemia.png

https://commons.wikimedia.org/wiki/File:Sucrose-inkscape.svg

https://commons.wikimedia.org/wiki/File:2511_A_Triglyceride_Molecule_%28a%29_Is_Broken_Down_Into_Monoglycerides_%28b%29.jpg

https://commons.wikimedia.org/wiki/File:404_Goblet_Cell_new.jpg

https://commons.wikimedia.org/wiki/File:Onion_root_cells.png

https://upload.wikimedia.org/wikipedia/commons/3/38/Adipocyte_types.jpg

https://commons.wikimedia.org/wiki/File:Glykogen.svg

https://commons.wikimedia.org/wiki/File:Cinnamycin.svg

www.ingramcontent.com/pod-product-compliance
Lightning Source LLC
Chambersburg PA
CBHW042049030726
47599CB00019B/2418